Planets tour

Best
Practice

Sun and the eight planets

Introduction

We live on just one planet. It is a solid home we have named Earth. Our planet revolves around the Sun. Earth never stops!

Our revolution of the Sun is known as an **orbit**. Two planets orbit nearer the Sun than we do. And five major planets are orbiting farther from the Sun than Earth orbits. This Best Practice book is your introduction to our Solar System with a fascinating journey to all major planets.

More moons orbiting outer planets have been found in recent times. And a change has been made to the number of major planets. For decades Pluto was regarded as one of nine major planets until it was reclassified as a "dwarf planet" in 2006.

So our Solar System has a total of eight major planets. From the Sun, these are: **Mercury, Venus, Earth, Mars, Jupiter, Saturn, Uranus, and Neptune.**

Enjoy the journey -
as far as your eye can see!

Contents

Planets tour

Solar System

Remarkable journey

Come along for a memorable journey which will take you out of your world! Your astronomic journey will soon begin from the middle of our Solar System – the Sun.

This epic adventure takes you through vast regions of space to eight major planets. All planets continually revolve along their path around the Sun. You will discover highlights and interesting facts about each planet.

Each one of these eight planets is unique and reveals a very compelling history. You will also discover how many moons each planet has continually in orbit around their primary planet. Every moon is also unique.

Before your journey begins let's gain an overview of our Solar System. It consists of the Sun and everything orbiting around it. This chapter uncovers intriguing facts about our nearest star – our one and only Sun.

4.6 billion years old

The formation of our Solar System began about 4.6 billion years ago! Following the collapse of a giant molecular cloud, gases and dust began combining to form the Sun, the planets, and all of the various other celestial bodies of our Solar System. These bodies include moons, asteroids, and comets. Our Solar System comprises the Sun and every astronomical object within the Sun's gravitational domain.

Solar

System

So our Solar System includes planets and their moons, asteroids, comets, dust, and gas. Planets, asteroids, and comets are all revolving around our Sun. Astronomy is the wonderful science of the sky and it is a fascinating study of all the celestial objects. True astronomy began in ancient Greece.

Sun is our only star

Most celestial bodies in our Solar System revolve around our Sun along nearly circular paths. All major planets and dwarf planets continually revolve, or orbit, on a path around our Sun in a counterclockwise direction. The word **"planet"** means **"wanderer."** Our Sun is the only star located within the boundary of our entire Solar System. And our Sun is actually the nearest star to planet Earth.

Solar

System

Planets shine simply because they reflect the rays of the Sun. All planets have no actual light of their own. Our Sun and our Solar System are a part of a much larger system, or galaxy, named the Milky Way. Our unfathomably huge Milky Way galaxy contains more than 200 billion stars!

Sun provides energy

A giant spinning ball of extremely hot gas, our Sun is fuelled by nuclear fusion reactions. We know light from our Sun heats our planet and makes life possible. Without energy from the Sun, we could not exist! Our Sun provides energy and light which is necessary to sustain life. The heat generated by sunlight is absorbed by our oceans and land masses. Sunlight keeps our surface at an average temperature of 14°C.

Solar

System

The Sun's inner core is the huge region where the energy and the nuclear fusion are produced. We know the Sun produces energy because we can see bright sunlight and we can feel the intense heat during a very hot day in summer. Our dazzling and distant Sun is roughly about 150 billion km away from us on planet Earth.

Largest body in Solar System

Our Sun's equatorial diameter measures 1.391 million km! This is almost 10 times longer than the diameter of the Solar System's largest planet Jupiter. Our Sun's **diameter** is about **109 times longer** than Earth's diameter. And since the quantity of an object's volume is proportional to its diameter cubed, the Sun's **volume** is consequently more than **one million** times **greater** than Earth's volume is!

Solar System

Our Sun's mass is 332776 times more than our planet's mass, despite the Sun having one million times more volume. This lower mass ratio differs from the respective volume ratio because our Earth has a considerably higher density than the Sun. Our massive Sun actually contains more than 99% of the total Solar System's mass!

Hydrogen is Sun's fuel

Our Sun is so huge you could fit one million planet Earths in our Sun's sphere! As the Sun is a spinning ball of hot gas, it does not have to rotate rigidly like solid planets and moons do. **Hydrogen**, the most abundant element in the universe, accounts for about **71%** majority of the Sun's mass. About 27% of its mass is helium. All of the other elements combined comprise less than 2% of the Sun's total mass.

Solar

System

Hydrogen acts as our Sun's **fuel**. Just a small amount of hydrogen can produce an immense amount of energy. There are incredibly vast reserves of hydrogen available to supply energy. These hydrogen reserves will constantly maintain the life of our Sun for an estimated five billion years or more into our future.

Sun's core is hottest part

The visible surface of the Sun is called the photosphere. This photosphere is where all the incredible energy, produced in the Sun's core, can finally escape the star. It is here the Sun's radiation is detected as visible light. From here amazing energy flies off into space. The surface of the Sun reaches temperatures of around 5000°C to 6000°C. However, the solar atmosphere is a hotter place to be than the photosphere is.

Solar

System

But the interior **core** of the Sun is also so much hotter than the photosphere is. The Sun's core is where all the work gets done and is where the nuclear fusion occurs. This core extends from the Sun's middle to roughly 25% of the total solar radius. The staggering high temperature of the core is estimated to be around 15 million° C!

Sun has equator & two poles

Our Sun continually rotates on its own axis. But the Sun as a whole body has no single period of rotation. The different portions of the Sun's surface perform their complete rotations in different time periods. The equatorial regions of the Sun's surface are rotating quite rapidly compared to the slower rotation occurring on other areas of the solar surface which are located away from its equator.

Solar
System

The Sun's period of rotation will vary somewhat with latitude. At the higher latitudes far from the equator the Sun rotates slower. It is estimated the Sun's rotation period varies from just 25 days at its equator to a much slower spin of 35 days at each of the Sun's two poles.

Planets follow elliptical path

Each planet's continual revolution of the Sun is not quite a perfectly circular path. This deviation from a precisely circular path is known as an "**elliptical**" path. This means there are times during each planet's elliptical orbit where the planet is orbiting marginally nearer to the Sun. At other times, the planet is somewhat farther from the Sun during each planet's less than perfectly circular shaped orbit.

Solar System

There are four inner planets and four major outer planets. They all follow their own elliptical orbit around the Sun. The inner planets orbiting nearer the Sun are Mercury, Venus, Earth, and Mars. Orbiting farther from the Sun are giant planets Jupiter, Saturn, Uranus, and Neptune.

Laws of planetary motion

During the early 17th century Johannes Kepler, a German mathematician, wrote his three laws of planetary motion. All of his three laws have still remained relevant in our Solar System today. And the three Laws of Planetary Motion form an essential part of the foundation of modern astronomy. The **first law** of planetary motion states the orbit of every planet follows an elliptical path around the Sun.

Solar

System

The second law states a line joining a planet and the Sun sweeps out equal areas in equal time. And the third law of planetary motion gives a mathematical link between a planet's orbital period and its distance from the Sun, so a Solar System scale model can be drawn.

Planets behave differently

The very different properties of the eight major planets are a direct result of each planet's proximity to the Sun. For planets orbiting on a path nearer the Sun, the higher extreme temperatures tend to drive away the light gases such as hydrogen and helium. This naturally means these hotter solid inner planets have therefore been left with the much heavier elements and the much denser elements.

Solar

System

Each planet has a different period of orbit, known as a "**year,**" around the Sun. Also, each planet has a different period of time for rotating fully on its axis. And each planet's rotational axis has its own angle of **axial** tilt. This angle is measured in degrees to the plane of the planet's orbital path.

MESSENGER spacecraft sent to Mercury

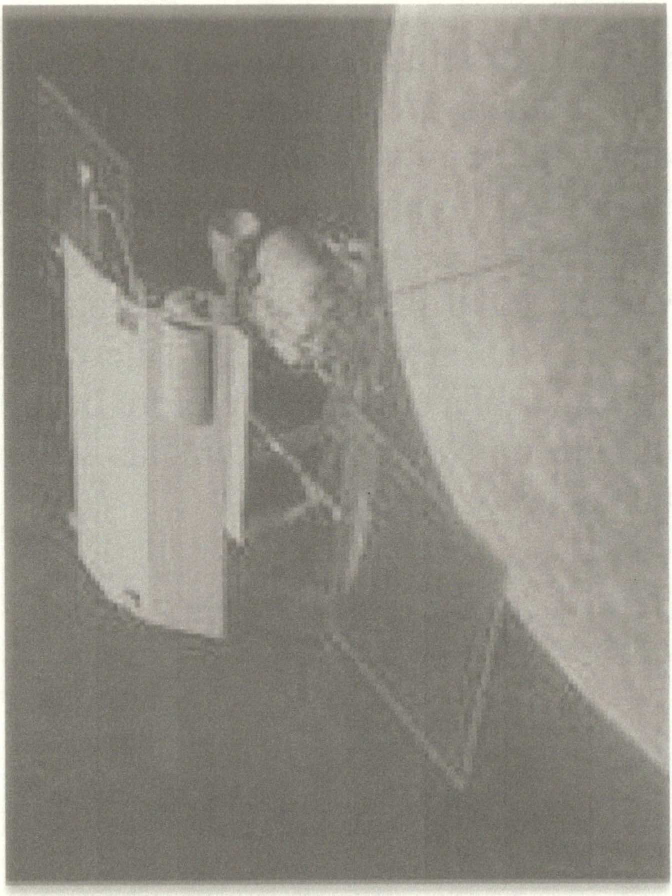

During MESSENGER's first flyby of Mercury in 2008, it became the second mission after Mariner 10 to reach Mercury. MESSENGER flew by Earth once, Venus twice, and Mercury three times. The main mission was successfully achieved in 2012.

Mercury

Journey begins

The famous crooner Frank Sinatra sang the popular jazz lyric, and more recently so did Canadian singer Michael Bublé; "Come fly with me…… let's fly…... let's fly away."

Most of your journey will be through vast areas of empty space. There are colossal distances of nothing at all! But there are planets, moons, and asteroids to visit. Let's find some planets to explore!

Planets of our Solar System all orbit the Sun, so this journey begins from there and moves outward. You will visit all major planets, but your **first** destination is **Mercury**. Space missions to Mercury include unmanned flybys in 1974/1975 and also in 2008/2009.

This interesting and compact planet is unique. It is the first planet orbiting our Sun. In the next twelve pages, we observe the highlights of the small but major planet named MERCURY:

Smallest planet

Mercury is a small and rocky planet measuring only 4875 km in diameter. Mercury is the nearest planet to the Sun and it has no moons and no natural satellites. Compact Mercury also contains no substantial atmosphere. And Mercury is physically even smaller than each of the two largest moons in our Solar System! The largest moons are Ganymede (orbiting Jupiter) and Titan (orbiting Saturn).

Mercury

The total mass of planet Mercury weighs only about 6% of the total mass of our Earth. Mercury's diameter is only 38% as long as Earth's diameter. And Mercury's gravity measures only about 38% as strong as Earth's gravity. Mercury is just slightly bigger than our Earth's Moon.

Orbit is the fastest

Mercury moves along its orbital path through space at a tremendous average speed of 47.9 kilometers per second! It takes **light**, the fastest moving object in the entire universe, about 3 minutes to reach the planet of Mercury after the light actually departs the Sun. And light will take about 8.3 minutes to travel all the way to our planet Earth from the moment the light is beamed from the Sun!

Mercury

Mercury revolves around the Sun very quickly, taking only 88 days to complete an orbit. Its orbital period is the **fastest** of any planet. But Mercury spins on its axis slowly, taking 58.6 days to fully rotate. Mercury has been named after a Roman fleet-footed messenger of the gods.

Covered by plains

The vast majority of Mercury's surface is covered by plains. There are two very distinct plains regions seen on Mercury. One region has **smooth** plains and the other region has **inter-crater** plains. The region which contains the smooth plains appears to be geologically much younger and it also contains fewer craters. In some of these areas patches of smooth lava can be seen to have filled part of the craters.

Mercury

The inter-crater plains are gently rolling or hilly plains. They exist in the regions between larger craters. These inter-crater plains were formed as lava flows buried the older terrain. The surface of Mercury is no longer active and it remains much the same as it has been for millions of years.

Hundreds of craters

Mercury's surface appears much like the surface of Earth's Moon and includes craters, mountains, valleys, and ridges. The numerous craters range in size from only 100 meters to 1500 km in diameter. More than 760 craters have been identified. Nearly 300 of the largest craters located on Mercury have been named. Most large craters are named after our famous writers, artists, and composers such as Mark Twain, John Lennon, Schubert, and Michelangelo.

Mercury

One striking feature on Mercury is a huge impact crater measuring about 1500 km across. This crater is named the Caloris Basin, and is surrounded by plains which look unusually bright. This is possibly because the huge impact which created the ringed basin excavated deep material.

Thin atmosphere

Mercury is dry, extremely hot, and is **almost airless**. The undersized planet is too small for its gravity to retain any significant atmosphere over long periods of time. Its atmosphere is essentially a **vacuum**. The very weak atmosphere of Mercury contains mainly sodium, but its atmosphere also comprises hydrogen, helium, oxygen, potassium, and calcium. The thin air does not trap heat very well.

Due to the immense heat on Mercury, the very thin atmosphere is blasted off its surface by the pressures of sunlight and the solar wind. Mercury's atmosphere quickly **escapes** into space and this entire thin atmosphere of Mercury is constantly being replenished with gases added to it.

Highs and lows

Mercury experiences some very extreme variations in temperature. The side of Mercury which currently faces the Sun reaches a blistering high of **452°C**. But the "dark" side of Mercury, which is not facing the Sun, experiences an absolutely frigid temperature which falls to a low of **minus 170°C!** So the extraordinary difference between the maximum temperature and the minimum temperature is about 622°C.

Mercury

Mercury orbits at a distance only about 30% of the distance from Earth to the Sun. The sunlight shining on Mercury's surface is 6.5 times more intense than it is on Earth due to Mercury's much closer proximity to the Sun. This extremely fierce intensity of sunlight would definitely be terribly hostile to any life on the planet.

No seasons

Mercury is not tilted at all with respect to its own orbit and this small planet therefore has no seasons. This is very unlike our Earth which has its rotational axis tilted at an angle of 23° to its orbital plane. Of all the planets in the Solar System, Mercury has the smallest angle measurement of **axial tilt**. Any planet which has no amount of axial tilt relative to the planet's orbit will naturally have no seasons at all.

Mercury

A planet's axial tilt is the angle between a planet's rotational axis and the line perpendicular to its orbital plane. Mercury's axis is directly perpendicular to its motion which means it is not tilted at all, giving Mercury no seasons. All of the other Solar System planets have seasons.

Three spins per two orbits

Mercury's period of rotation is 58.6 days and its orbital period is 88 days. Mercury spins in a counterclockwise direction from east to west. And Mercury also orbits the Sun in a counterclockwise direction, as do all of the other planets in the Solar System. Mercury will rotate fully on its axis three times during two of its complete orbits of the Sun. This 3:2 relationship is referred to as a **3:2 spin-orbit resonance**.

Mercury

For years it was thought Mercury was tidally locked with the Sun. The belief was Mercury had one rotation for each orbit. If so, Mercury would always have the same face towards the Sun. But the radar observations in 1965 proved Mercury has **three spins** for every **two orbits**.

Very dense

Mercury has the second highest density measure of all the planets in our Solar System. The very high density of Mercury measures just slightly less than our Earth's density. The metallic inner core of Mercury is extremely rich in iron. This small planet's inner core is actually larger than the entire sphere of our Earth's Moon. Mercury's core represents a significant 42% of the planet's total volume.

Mercury

Mercury's core is estimated to account for 75% of the planet's radius. The main reason for Mercury's very high measure of density is due to its metallic inner core being quite large relative to the small size of the planet. Mercury comprises about 70% metal and approximately 30% silicate.

Most eccentric orbit

Of all the major planets, Mercury has the highest deviation from a circular orbit. This deviation from a perfectly circular orbit is referred to as "**eccentricity.**" Mercury's eccentricity is measured at 0.206. This is more than double the eccentricity of any other major planet. Mercury's **elliptical** orbit of the Sun **ranges** dramatically between its nearest point to the Sun and its farthest point from the Sun.

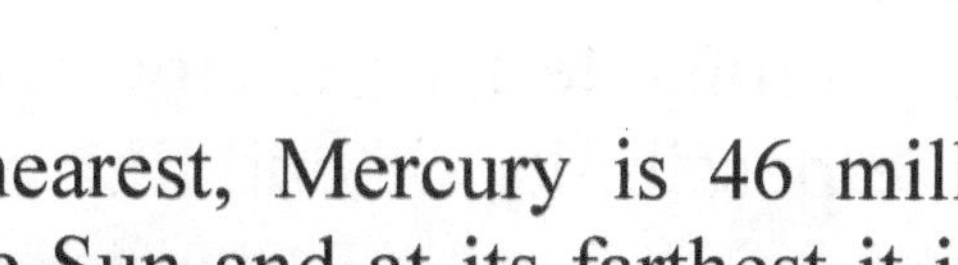

At its nearest, Mercury is 46 million km from the Sun and at its farthest it is nearly 70 million km from the Sun. Pluto has a more eccentric orbit than Mercury has but in 2006 Pluto was reclassified as a "dwarf planet", so Mercury retains the title of the most eccentric planet in our Solar System.

No probe has landed

So far, Mercury has never had any human visitors. This is because the ability to travel into outer space has only been something humanity has been capable of for the past sixty years. Although our technology is still rather primitive, the ongoing debate about whether humans should travel to other planets continues.

Mercury

Most people are fixated on the Moon and Mars as a probable target. However, there is also the possibility of building a balloon cloud city on Venus and a crater pod city on Mercury to study the Sun. The rise of well-funded spaceship corporations like SpaceX may open doors to sending humans where they have never gone before, including to the first planet, Mercury.

Lot of chromium

We use chromium to make stainless steel and shiny plating for classic cars. Plating spacecraft with stainless steel protects them from corrosion when confronting seawater as they land in oceans. SpaceX makes their Falcon and Starship rockets from stainless steel, which includes chromium. In future, Mercury may be ideal for mining the element because it has a lot of chromium.

Mercury

The percentage of chromium on Mercury is the greatest in its heavy core. Chromium is resistant to corrosion, a concern with structures and vehicles exposed to outdoor weathering and toxic acids. The average abundance of chromium on Mercury's surface, according to the MESSENGER space probe's data, is 200 parts per million.

Venus Express

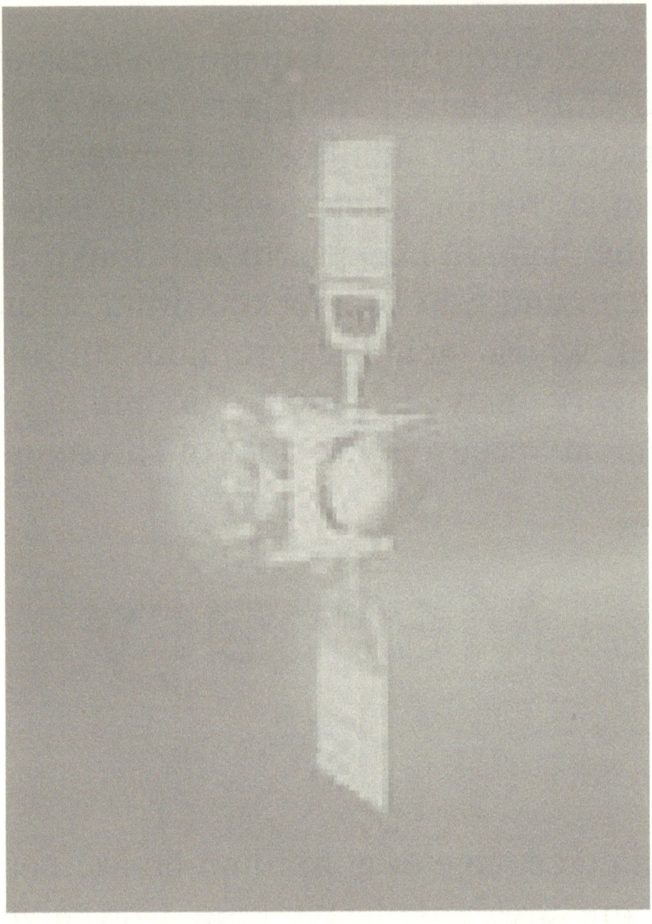

Venus Express was the European Space Agency's first exploration mission to Venus. It studied the Venus atmosphere and the planet's plasma environment. Meteorological maps were also created. The mission continued from 2005 to 2014.

Venus

This sister is hot!

Departing Mercury, your voyage through space now moves farther from the Sun. Your next destination is the second closest planet to the Sun. This nearby planet is Venus.

Planets orbiting nearer to the Sun than Earth's orbit of the Sun are referred to as "inferior" planets. There are only two inferior planets. Mercury and Venus both orbit the Sun inside Earth's orbit.

The orbital path of Venus is nearer to Earth's path around the Sun than any other planet's path is. Until the 1960s Venus was often referred to as Earth's "twin sister". But Venus is different to Earth in so many ways.

You will now learn about some of these key differences between Venus and Earth. Also contemplate the following features of the hot planet named VENUS:

Hottest planet

Venus is an intensely hot planet but it never faces a sunny day! The Venus atmosphere is extremely dense and it is so laden with clouds it hides the surface of Venus permanently. The really torrid surface temperature soars to an astounding 500° Celsius! The very thick clouds permanently covering Venus create a **greenhouse effect** and this causes the planet's surface to be unbearably hot.

Venus

The greenhouse effect on Venus is a bit like our global warming but on a crazier scale. The incredible heat which typifies the surface of Venus makes it totally unfit for life. Venus is hotter than Mercury is, despite Venus orbiting twice as far from the Sun than the nearer Mercury orbits.

Spins slowly backwards

From Earth, Venus appears **upside-down** as it spins from east to west. All planets orbit in a counterclockwise direction and most planets also rotate counterclockwise. But Venus **rotates** in a **clockwise** direction. If you were on Venus, the Sun appears to be rising in the west and later the Sun appears to set in the east. This unusual sunrise and sunset is the opposite of what we observe on Earth.

Venus

But an observer on Venus would not see anything overhead other than a sky laden with **clouds**. So sunrise to the west and sunset to the east from a Venus view is a purely academic consideration. This is simply because the Sun would not be visible beyond the incredibly thick clouds!

Brightest planet

Venus is the brightest object in our sky besides our Sun and our Moon. It is in fact the most reflective planet. Venus is known as the "**morning star**" because it can be seen at sunrise and it appears in our east. Venus has also been referred to as the "**evening star**" because it can be seen at sunset and it appears in our west. Venus cannot be seen at all in our sky during the middle section of our night.

Venus

Of all the planets, Venus is the closest to our Earth in physical size. The diameter of Venus is about 95% of Earth's diameter and the mass of Venus is about 82% of Earth's mass. Venus was named after the Roman goddess of love and beauty. The Greek equivalent of Venus is Aphrodite.

Mostly plains

The surface of Venus is an interesting combination of plains, highlands, lowlands, and mountains. The vast rolling plains cover 65% of the surface and the Venus highlands account for only about 15%. The desolate surface of Venus is very barren with much of it covered with old lava flows. Many of the volcanic features, created on the Venus surface billions of years ago, are still visible today.

Venus has two major mountain ranges:
1) In the **northern** hemisphere is **Ishtar Terra** which is a similar size to Australia. The highest mountain on Ishtar Terra's range is Maxwell Montes, 12 km high.
2) In the southern hemisphere of Venus is Aphrodite Terra, similar in size to Africa.

Hidden by clouds

The bright layers of clouds above Venus all comprise sulphuric acid. These clouds move at a speed of up to 350 km per hour. They actually rotate around Venus at a speed about 60 times faster than Venus itself rotates! These bright clouds take only four Earth days to fully circle Venus! But the planet of Venus itself rotates slowly on its own rotational axis in approximately the equivalent period of 243 Earth days.

Venus

These stacks of bright clouds distribute a lot of heat around the entire planet, so the "night" side of Venus remains really hot also. The Venus clouds rain sulphuric acid which is an excessively acidic rain. But this awfully corrosive precipitation does not actually reach the planet's surface.

Most circular orbit

Of all the planets, Venus has the lowest measure of any deviation from a strictly circular orbital path. This deviation is known as the **eccentricity** of its path. The orbital eccentricity of Venus is measured at a tiny 0.007, which is the lowest orbital eccentricity of any planet. Neptune has the second most nearly circular elliptical orbit with a low orbital eccentricity which has a measurement of just 0.009.

Venus

The shortest distance Venus will get from the Sun is 107.5 million km. This nearest point, on a planet's orbital path, to the Sun is referred to as a planet's **perihelion**. And the longest distance Venus will get from the Sun is a fairly similar distance of 108.9 million km.

Dense atmosphere

Carbon dioxide is absolutely the major constituent of the upper atmosphere of Venus. The thick atmosphere of Venus contains 96.5% carbon dioxide and only 3.5% nitrogen. And the weight of this very dense atmosphere is about 90 times greater than the weight of our Earth's atmosphere. The extremely heavy atmosphere of Venus is also very much denser than the atmospheres of both Mercury and Mars.

Venus

You would feel immense **pressure** if you were ever able to stand on the surface of Venus. This crushing pressure, or force, would be fairly similar to the enormous pressure you would experience if you were completely submerged at the bottom of one kilometer of water in a deep ocean!

Most volcanoes

The rocky surface of Venus appears to have been created through widespread volcanic activity. There are more than **1600 volcanoes** on Venus and most of these are considered to be very long extinct. The highest volcano located on Venus has been named Maat Mons. This towering and impressive volcano has a height of 4.9 km from its base to its peak. It is fairly similar in height to Earth's Mount Everest.

Venus' surface is about 90% basalt, which is a volcanic rock. And 80% of the surface area has a mosaic of volcanic lava plains. This proves very **extensive volcanic activity** has performed a major role in shaping the surface of Venus. No other planet has as many volcanoes as Venus.

Crust is single plate

Although the surface of Venus can never be observed visually, due to the planet being hidden by thick clouds, the intriguing surface has been mapped in considerable detail. The Venetian crust displays tectonic activity with extensive faulting and volcanism. The crust of Earth has a jigsaw assembly of several tectonic plates. But the **upper crust** of Venus is so different to Earth as it is effectively only a **single plate**.

Venus

Sprawling **mountain plateau** on Venus are the result of hundreds of millions of years of intensive volcanism occurring at very specific points in a static crust. This extensive volcanism has enabled vast amounts of material to extrude on to the surface and sprawl across the terrain.

Year shorter than day!

The orbital period for Venus to revolve fully around the Sun (a Venus "**year**") is equivalent to 224.7 Earth days. And the planet's complete axial rotation period (a Venus "**day**") is the equivalent of 243 Earth days. Venus rotates on its own axis extremely slowly. This means we have the really unusual situation occurring where a Venus "**day**" will actually continue **longer** than a Venus "**year**" will continue for!

Venus

The extremely long axial rotation period of **243 days** for Venus is by far the **longest** axial **rotation** period of any major planet. This really slow east-to-west retrograde spin on its rotational axis gives Venus has the title of the slowest spinning planet. And Mercury is the second slowest spinner.

First probe on another planet

When Soviet spacecraft *Venera 7* made a soft landing on Venus, it became the first spacecraft to land on the surface of another planet. It was the first probe to send data from Venus to Earth. Launched in 1970, *Venera 7* entered the atmosphere of Venus on December 15, 1970, and then **landed on planet Venus** the same day.

Venus

The Venera (Russian for Venus) probes (1961-1984) were developed to gain more data from Venus. Ten probes landed on Venus and transmitted data from its **surface**. And 13 Venera probes transmitted data to us from the **atmosphere** of Venus. In 1967, Venera probes created history by becoming the first manufactured devices to enter another planet's atmosphere.

What the future holds

Researchers want to better understand every planet in our solar system. Efforts in the late 20th century showed that Venus is a difficult planet to observe remotely from the surface, but with new technologies and a better understanding comes innovative exploration ideas. A lot of these new ideas have a common theme, which is exploring Venus from within the clouds.

Venus

As Venus has more favorable conditions in the clouds, with wind speeds that allow an object to travel around the planet much faster than it rotates, scientists are looking to introduce airships. By utilizing solar and wind power, and the help of buoyancy, robotic missions could become a feature of Venus in the foreseeable future.

Earth and Moon from space

Earth

Special planet

Departing Venus, you now travel to a very familiar planet you cherish and know intimately. Yes - it is our wonderful Earth.

We know considerably more about the third closest planet to the Sun than we know about all other planets combined. Your life span on Earth is so brief when compared to the billions of years since Earth was created.

It is essential we take care of this special planet. Not just for now, but for all of our future generations to follow after us. We are sharing the one planet and each of us has a genuine ongoing responsibility to maintain Earth's delicate and fragile balance. We are custodians of a planet!

Let's observe some of the interesting facts about our home planet and also learn about the unique characteristics of our really special planet we have named EARTH:

Largest inner planet

The equatorial diameter of our planet is a modest 12756 km. So Earth is **larger** than all of the other three rocky **inner** planets Mercury, Venus, and Mars. These planets are also referred to as "terrestrial." But our Earth is very much **smaller** than all four of the **outer** planets Jupiter, Saturn, Uranus and Neptune. This means Earth is ranked as the fifth largest in both size and weight of the major planets in our Solar System.

Earth

Earth is not a perfectly shaped sphere due to a **bulge** at the equator. This bulge results from the continual rotation of our Earth on its axis. This equatorial bulge has caused the diameter of our equator to be roughly 43 km more than our polar diameter which is between North Pole and the South Pole.

Orbit takes 365.256 days

Earth orbits the Sun at a whirlwind average speed of about 30 km per second! Our orbit of the Sun takes exactly 365.256 days to complete one revolution. Most of our calendar years contain 365 days, but every four years we have a **leap year** which adds an extra day to our year thus making a total of 366 days. This leap year accommodates for the extra fraction (.256) of a day which forms part of our total period of orbit.

Earth

Earth's elliptical orbit means we are nearest to the Sun between January 2 and January 5 each year. And we are farthest from the Sun between July 4 and July 7. Over one year, our **southern** hemisphere gets just a little **more** energy from the Sun than our northern hemisphere receives.

Three layers

Earth's innermost layer of its three solid layers is known as the **core**. Our central core is clearly the hottest part of the whole planet, and the core is comprised of molten nickel and iron. The movement of this iron-rich inner core generates Earth's magnetic field. The really intense heat which comes from deep within our inner core causes the material in both the Earth's outer core and the Earth's mantle to be moved around.

Earth

Our second layer is known as the **mantle**. It consists of solid, hot rock. The mantle is clearly the thickest of the three layers as it represents about 84% of Earth's volume. Our uppermost layer is known as the **crust**. It is this thin crust upon which we all live and walk. Our crust consists of igneous, metamorphic, and sedimentary rocks.

Bound by blanket of air

Our planet is surrounded by a blanket of air which is known as our **atmosphere**. The overriding majority (99%) of our total atmosphere is contained in just the two lowest layers: our **troposphere** and our **stratosphere**. The lower troposphere is the layer where we all live and this is where our weather occurs. This troposphere layer is the region contained within about 12 km above the Earth's surface.

Earth

Earth's atmosphere is a covering of gases surrounding our solid planet. This blanket of air is retained by Earth's gravity. Our air actually gets **thinner** as the altitude gets higher. Earth's atmosphere is comprised of about 78% nitrogen and 21% oxygen. It also contains argon and other trace gases.

Rotates in 23.9345 hours

Earth is continually rotating towards the east in a counterclockwise direction. Earth is spinning relentlessly around its own axis. It is this spinning of our Earth which causes our Sun to appear to be moving from east to west across our sky. The length of one "**solar day**" is exactly 24 hours. But Earth's actual period for one complete rotation on its axis is an exact 23 hours, 56 minutes, and 4 seconds.

Earth

As Earth is rotating, our planet keeps moving around the Sun in our orbit's small **interval** from one day to the next. Thus our 24-hour solar day is just slightly longer than our actual period of rotation because Earth will rotate just a fraction more than one full turn in space each and every day.

We have four seasons

Earth's four seasons are determined by our two **solstices** and our two **equinoxes**. Our two solstices are the points in our orbit of the maximum axial tilt **toward** the Sun (summer solstice), and also the maximum axial tilt **away** from the Sun (winter solstice). And our two equinoxes occur when the direction of Earth's axial **tilt** and the Earth's **direction** of its orbital path are **perpendicular** to each other.

Earth

Our northern hemisphere winter solstice is near December 21 and summer solstice is near June 21. Equinoxes are near March 20 in spring, and September 22 in autumn. The four seasons of our southern hemisphere are reversed as its axial tilt opposes the northern hemisphere's tilt.

We are all on one planet

Earth is so very special! Earth is the only planet we know of to support life. A planet has to be able to meet certain conditions so it can support life. The most important life-supporting condition is for the planet to lie in the **habitable zone**. Basically this habitable zone is a region around a star, or sun, in which a planet has the ideal temperatures which will permit the availability of liquid water.

Fortunately, Earth meets life-supporting conditions. These conditions include that a planet is solid to allow for a liquid-solid interface. This enhances the exchange between molecules. And Earth is located at just the right distance from our Sun to allow for liquid **water** to be present.

Our land masses

Our continents are the great land masses on planet Earth. There are **seven** large continents on Earth: Antarctica, Africa, Asia, Australia, Europe, North America, and South America. And thousands of islands are also included in the total land content of these seven continents. The continent which comprises the largest amount of total land area is Asia. The world's smallest continent is Australia.

Earth

Earth's second largest continent is Africa, which has the most countries on its land mass. Asia is home to more than half of Earth's population. Total **land** mass of all continents combined is over 148 million square km. This land mass area accounts for only **29.1%** of the total surface area.

Covered mostly by water

From outer space, Earth appears as a blue world. Our planet is often affectionately referred to as the **"blue marble."** Earth appears distinctly blue because about 70% of the Earth's surface is covered with water. An overwhelming 97% of all this surface water is in our salt water oceans. So this means only 3% of our surface water quantity is available for people to use as a drinkable fresh water supply.

Earth

Water is vital for all life. Clean drinking water is essential to humans. Our water is continually moving through its natural **cycle** of evaporation, precipitation, and runoff. Most of the water runoff reaches one of our seas. About 70% of fresh water is consumed by our agriculture.

Modern wonders

During the last 100 years, **people** have made many great changes to our planet. Travel is much quicker, communication has improved, cities are larger, and our population has grown considerably. We have witnessed monumental engineering and construction feats. Our Earth has so many natural wonders but a very short list of **seven** man-made **modern wonders** of our world includes the following:

Earth

1) Empire State Building - New York
2) Itaipu Dam - Brazil & Paraguay
3) CN Tower - Toronto
4) Panama Canal
5) Channel Tunnel: Paris ↔ London
6) Golden Gate Bridge - San Francisco
7) North Sea Protection Works

First human is space

Yuri Gagarin became the first person in outer space when he flew aboard the *Vostok 1* spacecraft in **1961**. Gagarin made one orbit of Earth on the historic flight lasting 1 hour and 48 minutes. Gagarin was only 1.57 meters tall, which was a distinct advantage in the small *Vostok 1* cockpit. After his notable flight on April 12, 1961, Gagarin became a worldwide celebrity.

Earth

Gagarin was the first man to see our Earth is definitely **round**, the planet is mostly **water** and indeed magnificent. First words from a man in space were: "*I see Earth, it is so beautiful!*" Yuri also said, "*The Earth is blue... how wonderful... it is amazing.*" A crater on the far side of the Moon is named after Gagarin.

Largest artificial satellite

At the top of a planet's atmosphere is the region called **low orbit**. This area is within the influence of a planet's gravitational field. Earth's low orbit is within about 2000 km of our surface. Low orbit is the region where many artificial satellites will hold their orbit. The **International Space Station**, the largest artificial **satellite** to orbit Earth, is in our low orbit region.

Earth

All human space flights to date were made in Earth's low orbit region, except for lunar flights of the **Apollo** program. Our low orbit environment has become congested with space debris. There are more than 8500 pieces of space debris identified as being larger than 10 cm. There are also millions of smaller debris pieces in space!

Unmanned missions to Mars

Since 1965, voyages to Mars have revealed a world strangely familiar to ours. But conditions on Mars vary wildly from what we know on our own planet.

Mars

Last inner planet

Your journey now departs our wonderful Earth and takes you farther again from the Sun. Get ready to enter a cold desert world. Let's explore the last inner planet: Mars.

Planets with an orbit farther from the Sun than Earth's orbital path is are referred to as "superior" planets. This next stop on your amazing tour makes Mars your first visit to a "superior" planet.

Our missions to Mars included failed attempts to obtain images of Mars dating back to 1960. Unmanned spacecraft sent to Mars have achieved many successful flybys and orbits since 1965. More recently, Exploration Rovers have landed on Mars.

Are men from Mars and women from Venus? Not likely! But let's now look at the interesting and unique aspects regarding the arid and mystical planet of MARS:

Red planet

Mars is often referred to as the **red planet** because of its distinctly red and **rusty** appearance. The rust-colored portions we can see on the surface of Mars are considered to be much like desert regions. The iron oxide which is very prevalent in the layers of surface dust covering Mars gives the planet its bloody tinge and very recognizable reddish hue. This surface layer of fine red dust is quite shallow.

Mars

Mars can actually be seen from planet Earth with the naked eye. The **apparent magnitude** of Mars in our sky is only surpassed by the brightness of our Sun, our Earth's Moon, the planet Venus, and sometimes by Jupiter. The red planet is named after the Roman god of war; Mars.

Smaller than Earth

Mars is not really enormous when it is compared to other major planets. The diameter of Mars is just 6794 km across. And the gravity on Mars measures only 38% of the gravity we experience on Earth. Mars is the **second smallest** major planet in our Solar System with only diminutive Mercury being a physically smaller size. The total surface area of Mars measures nearly 145 million square km.

Mars

The diameter of Mars is only about **half** of Earth's diameter. Mars has a lower density than our planet's density. Mars also has about 15% of Earth's volume and only about 11% of our mass. The total surface area of Mars is just a little less than the total area of Earth's dry land regions.

Polar caps

A **polar cap** is the very **highest latitude** region of a planet or a natural satellite. Mars has two brilliant permanent polar ice caps. Whilst Earth's two polar caps are mainly composed of water ice, the polar caps on Mars are a **mixture** of solid phase **carbon dioxide** and **water ice**. The two polar ice caps on Mars will change their appearance quite significantly with the change of the planet's seasons.

Mars

At these polar caps carbon dioxide ice sublimes in summer which uncovers a surface of rocks. This carbon dioxide ice then reforms in winter. During both the northern pole and southern pole winters on Mars the respective winter **pole** is in continual **darkness**. Average temperature on Mars is an unfriendly minus 63°C.

Highest peak in Solar System

The Tharsis region of Mars is dominated by four massive shield volcanoes. Located on the vast plateau of Tharsis are four of the largest known volcanoes named Arsia Mons, Ascraeus Mons, Olympus Mons, and Pavonis Mons. Mars is home to the **highest peak** in the entire Solar System. Olympus Mons (elevation 21,171 meters) is nearly three times taller than Earth's highest peak which is Mount Everest.

Mars

Olympus Mons is also the **largest** known **volcano** in the Solar System. It resembles Hawaiian shield volcanoes. The central edifice of Olympus Mons has a very flat summit caldera which rises to 24 km above the surrounding plain! The diameter of this towering volcano's base is 620 km.

Contrasting hemispheres

Mars has some of the most interesting terrain of the inner terrestrial planets. Some of its surface is quite spectacular. The northern hemisphere and the southern hemisphere appear quite contrasting to each other. The surface of the **southern** hemisphere is similar to the surface of our Moon. This southern hemisphere of Mars has the thicker crust and its older surface is mainly ancient, cratered highlands.

Mars

In stark contrast, the surface of the Mars **northern** hemisphere appears very much brighter. The vivid northern hemisphere consists mostly of plains which are much **younger** and are also lower in elevation. These lava-filled plains have had a more complex history of activity and change.

Seasons like Earth

Of all the planets, the four seasons on Mars (summer, autumn, winter, spring) are the **most** comparable to the seasons we have occurring on Earth. This similarity of the four seasons on Mars and Earth is directly due to the very similar amount of tilt of each planet's rotational axis. But all of the four seasons on Mars continue for a much longer period than each of the four seasons we experience on Earth.

Mars

Spring on the southern hemisphere of Mars (northern autumn season) goes for 146 days. The southern hemisphere's summer (northern winter) continues for 160 days. The southern autumn (northern spring) lasts for 199 days, and the southern hemisphere's winter season (northern summer) continues for 182 days on Mars.

Day 37 minutes longer!

Mars rotates fully on its axis once every 24.6 hours. This means a complete "day" on Mars is just 37 minutes longer than the 24-hour day we experience on Earth. Mars rotates fully on its axis 670 times per year. And Mars progresses along in its revolution around the Sun about one half of just one degree of the circular path during the planet's full rotation. Mars rotates at a reasonably similar rate to the rotation of our Earth on its axis.

Mars

If Mars is observed at the same time each night its rotation appears to lag behind a little more each night. This difference in our observation of Mars is only 9° of Martian longitude each day. This means a complete tour around Mars can be achieved in **six weeks** of observation!

Thin atmosphere

The atmosphere of Mars is very different to the atmosphere of our Earth. The main component of the atmosphere of Mars is carbon dioxide representing a huge **95%**. The other elements are 3% nitrogen, 1.6% noble gas argon and traces of oxygen, water, and methane. When compared to Earth the atmosphere of Mars is quite "**rarefied,**" which means the air surrounding Mars is very much thinner.

Mars

The remarkably thin atmosphere of Mars contains small, suspended **dust** particles, giving its sky an orange or light brown tinge. Atmospheric pressure measured on the surface of Mars is extremely low. The dusty atmosphere of Mars is only a tiny 0.5% as thick as Earth's atmosphere.

Largest canyon

The extremely large canyon named Valles Marineris has a length of more than 4,000 km. This canyon reaches depths of up to seven kilometers which is more than four times deeper than our Grand Canyon. The distance of this Valles Marineris canyon system is roughly equivalent to the length of Earth's European continent! This long canyon can be seen extending across nearly one **quarter** of the **circumference** of Mars.

Mars

When Mars is viewed from space, Valles Marineris is prominent and appears to cut a wide gash across the surface. The canyon is located just south of the Mars equator, on the south-eastern side of the Tharsis area. Valles Marineris is the **largest** known **canyon** in our Solar System, and it is also longer than any other known crevice.

Year takes 687 days

The average distance between the Sun and Mars is nearly 228 million km. But Mars gets as far away as 249 million km from the Sun. This most distant point from the Sun is known as a planet's **"aphelion."** Mars will take 687 Earth days to complete its orbit of the Sun. One "year" on Mars is the equivalent of about 1.88 Earth years. Mars has the second highest **eccentricity** of any major Solar System planet.

Mars

Only Mercury's orbit deviates more from a strictly circular path than the orbit of Mars does. Mars once had a more circular orbit. But for the last 35000 years the orbit of Mars has gradually become more eccentric. This change in eccentricity has been due to gravitational effects of the other planets.

Mars missions

Robotic spacecraft began observing Mars in the 1960s, with USA launching *Mariner 4* in 1964 and *Mariners 6* and *7* in 1969. Those early missions revealed Mars to be a barren world, without any signs of the life or civilizations. In 1971, *Mariner 9* orbited Mars, mapping about 80% of the planet, discovering its volcanoes and big canyons.

Mars

The Soviet Union also launched numerous Red Planet spacecraft in the 1960s and early 1970s, but most missions failed. *Mars 2* (1971) and *Mars 3* (1971) operated successfully but were unable to map the surface due to dust storms. NASA's *Viking 1* lander made the first successful landing on Mars in 1976. Its twin, *Viking 2*, landed six weeks later in a different Mars region.

Mars rovers

In 2003, Mars passed closer to Earth than it had on any occasion in the past 60,000 years. Also in 2003, NASA launched two golf-cart-sized rovers named Spirit and *Opportunity.* These two rovers explored different regions of the Martian surface after touching down in January 2004. Both rovers found many signs that water once flowed on the planet's surface.

Mars

Spirit and *Opportunity* were originally tasked with three-month surface missions, but both kept roving much longer. NASA didn't declare Spirit dead until 2011, and *Opportunity* was going strong until a dust storm hit in 2018. In 2008, a NASA lander *Phoenix* confirmed the presence of water ice in the subsurface, among other finds.

Juno spacecraft to Jupiter

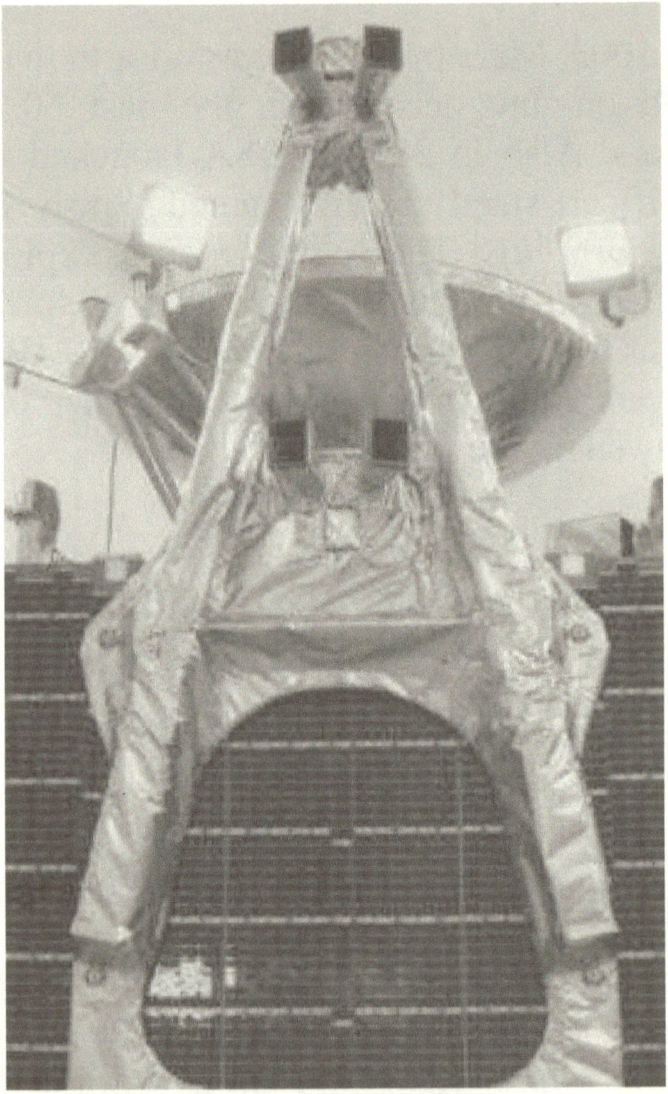

Juno will explore Jupiter's atmosphere. The principal goal is to understand the origin and evolution of giant Jupiter.

Jupiter

Across the Belt

Your fascinating journey to planets now transfers from the four inner "rocky" planets to the four outer "gaseous" planets.

This tour now moves across a huge region, or belt, of interplanetary debris located in between the orbits of Mars and Jupiter.

Asteroids and comets are continually orbiting the Sun within this incredibly large region known as the **Asteroid Belt**.

There are hundreds of thousands of small asteroids orbiting within this huge belt. But the asteroid material is thinly distributed in this vast area of space. There is one dwarf planet, named Ceres, in the Asteroid Belt.

After crossing the Asteroid Belt, you will encounter the fifth major planet from the Sun. Let's now closely inspect the really imposing giant planet JUPITER.

Largest planet

The extremely large equatorial diameter of Jupiter measures 143883 km. Jupiter is huge in comparison to all other planets in every regard. The diameter of Jupiter is roughly about eleven times **longer** than the diameter of Earth. Jupiter has a mass about 318 times **heavier** than the mass of our Earth! Jupiter is appreciably much larger physically and also so much heavier than any other planet in our Solar System.

Jupiter

Jupiter accommodates a whopping 71% of the total amount of planetary matter discovered in the Solar System! Jupiter's exceedingly massive volume is actually large enough to contain 1300 planets the size of Earth! The giant planet of Jupiter has quite appropriately been named after the **king** of the Roman gods.

Ball of gas and liquid

Jupiter does not have the similar type of conventional solid surface we are accustomed to on Earth. Jupiter is one of the four "gas giants" but it does contain a solid and rocky inner **core**. By mass, Jupiter's atmosphere has a composition of approximately 75% hydrogen and about 24% helium. And the remaining 1% of the Jupiter atmosphere contains small traces of ammonia, methane, and water vapour.

Jupiter

Jupiter's "surface" is dense red, brown, yellow, and white clouds grouped in lighter colored areas (zones), and darker regions (belts). Cloud zones and cloud belts circle Jupiter parallel to its equator. The clouds are about 50 km thick and under the clouds is a big layer of **liquids**.

Cold and hot!

Jupiter has a big range of temperatures. Its outermost layers are exposed to the near **absolute zero** temperature of space. To contrast these very cold conditions, Jupiter's inner core has a really high temperature due to the intense pressure of the whole planet bearing down on it. Descending deeper and deeper towards Jupiter's inner core, the temperature rises dramatically to about 35,000°C!

Jupiter

At Jupiter's cloud tops the temperature is an exceedingly cold minus 145°C! Its temperatures **increase** as you descend **deeper** into the huge planet. When the pressure of the Jupiter atmosphere gets to about ten times the pressure we have on Earth, Jupiter is approximately 20°C.

Small tilt

All planets have an imaginary axis they rotate around. The north and south poles of all planets are the points where the imaginary axis would poke through the planet's surface. Jupiter has a small axial tilt of only **3°** which means there is very little difference between all four of its seasons. Jupiter has vivid aurora, which are very powerful glowing energies, which form over its north pole and its south pole.

Jupiter

Radiation at the poles affects Jupiter's atmosphere. Earth is bitterly cold at its north and south poles but much hotter nearer the equator. As most of Jupiter's **heat** comes from its **inner core** instead of coming from sunlight, the poles of Jupiter are equally as warm as the equator is!

Orbit takes 11.86 years

The period for Jupiter's full revolution around the Sun is 11.86 Earth years. That's a total of 4,333 days! Jupiter orbits so very much farther from the Sun than our Earth orbits the Sun. Therefore, it will take Jupiter considerably much **longer** to complete an orbit than it takes Earth to complete its orbit of the Sun. The speed of planet Jupiter during its continual orbit of the Sun is about 13 km per second!

Jupiter

Just like all the other planets, Jupiter follows an **elliptical** path around the Sun. Jupiter gets as close as 741 million km from the Sun. At its most distant point, Jupiter is about 817 million km from the Sun. Average distance between Jupiter and the Sun is about 778 million km.

Fastest rotating planet

Jupiter rotates faster on its own axis than any other planet rotates in the entire Solar System. This really huge planet **rotates** on its own axis so **quickly** the Jupiter "days" are only about 9.8 Earth hours long! This extremely short Jupiter day is hard to comprehend. Especially when you consider the incredibly massive Jupiter is measuring more than the equivalent size of eleven Earth diameters in its width.

Jupiter

Due to Jupiter's extremely rapid rotation, centrifugal forces cause Jupiter's shape to be an **oblate spheroid**. Jupiter is not a perfect sphere as a bulge is visible at its equator. The long equatorial diameter of Jupiter is about 7% longer than the polar diameter between north and south poles.

Faint rings

The faint **rings** orbiting Jupiter were the third set of planetary rings discovered in the Solar System. Jupiter’s rings were found after discoveries of the rings around Saturn, and rings around Uranus. Rings orbiting Jupiter are so much fainter than those seen around Saturn. The faint rings of Jupiter have probably been formed from material ejected by moons when they are struck by meteorites.

Jupiter

The faint rings continually orbiting Jupiter were first observed from Earth in 1979. Jupiter’s ring system comprises four main sections: one thick inner **halo ring**; one relatively bright and extremely thin **main ring**; and also two wider thick and really faint outer **gossamer rings**.

Enormous magnetic field

Our Earth is considered to have a strong **magnetic field.** However, the magnetic field at the tops of Jupiter's clouds is about ten times **stronger** than our Earth's magnetic field. This exceptionally broad magnetic field of Jupiter also contains twenty thousand times **more energy** than Earth's magnetic field contains! Jupiter has by far the most powerful magnetic field of any planet in our Solar System.

Jupiter

The intensity and complexity of Jupiter's enormous magnetic field is due to the planet's extremely rapid rotation and its substantial metallic hydrogen core. The staggering region of space dominated by the magnetic field of Jupiter is unbelievably enormous. This colossal region is known as its **magnetosphere**.

Powerful winds rule

Jupiter's **powerful winds** are so incredibly different from the much milder winds we experience here on Earth. Jupiter's vigorous and very energetic winds are relentlessly circling the huge planet. These winds have changed little over the last 300 years scientists studied them. The massive east-west winds which stream in Jupiter's equatorial region reach colossal speeds of about 550 km per hour.

Jupiter

Jupiter's wind speed is **twice** as rapid as the destructive winds generated by strong hurricanes on Earth. Jupiter's storms are often accompanied by **lightning strikes**. The vivid colors seen in storms probably result from the subtle chemical reactions of trace elements in the atmosphere.

Largest storm

The **Great Red Spot** of Jupiter is a large storm discovered in 1664. The really long diameter of the Great Red Spot is more than double the entire diameter of our planet Earth! The very eye-catching Great Red Spot storm has always been one of Jupiter's most **easily recognizable features**. This impressive and prominent storm is about 28,000 km in overall length and is about 14,000 km in width.

Jupiter

The Great Red Spot is a high pressure, or anti-cyclonic, persistent storm eternally circulating in the atmosphere of Jupiter's southern hemisphere. It takes about **six days** to circulate in a counterclockwise direction. The Great Red Spot is the largest known vortex in our Solar System.

Vivid colors

Jupiter's appearance is a lovely tapestry of colorful stripes and spots. Cloud bands encircle the planet, and cyclonic storms dot it from pole to pole. The gas planet likely has three distinct cloud layers in its "skies" that, taken together, span about 71 km. The top cloud is probably made of ammonia ice, while the middle layer is likely made of ammonium hydrosulphide crystals.

Jupiter

The innermost layer may be made of water ice and vapor. Jupiter's fast rotation creates strong jet streams, separating its clouds into dark belts and bright zones. Data from the spacecraft *Juno* shows that Jupiter’s cyclones are warmer on top and colder at the bottom. The planet’s anticyclones are colder on top and warmer at the bottom.

First encounter with Jupiter

NASA launched the space probe *Pioneer 10* in 1972. It was aboard an Atlas-Centaur rocket. Thid was the fastest spacecraft to leave Earth back then. *Pioneer 10* was fast enough to pass the Moon in 11 hours! It crossed the orbit of Mars just 12 weeks after leaving Earth. *Pioneer 10* was the first spacecraft to travel through the **Asteroid Belt**, and first o visit any outer planet.

Jupiter

In 1973, *Pioneer 10* went past the giant Jupiter. It was the first probe to make direct observations of **Jupiter** and to obtain close-up images. *Pioneer 10* located its magnetic field and established the planet is predominantly liquid. *Pioneer 10* was the first man-made object to **leave the Solar System** and travel into **interstellar space**.

Saturn from *Cassini* probe

Saturn

Glorious rings

Your astronomic adventure now moves from the "gas giant" Jupiter and ventures farther yet again from the Sun to greet another giant planet named Saturn.

The temperatures are clearly dropping much lower now as your intrepid journey proceeds away from the Sun. You are now nearing the sixth major planet orbiting the Sun.

Saturn is the most distant planet we can view from Earth without the aid of a telescope. Unmanned spacecraft have sent back many astonishing images of Saturn and its rings. This remarkable ringed planet is regarded by many observers as one of the most beautiful objects seen anywhere in the Solar System.

Awe-inspiring images show profoundly well-defined rings. Let's now appreciate this beautiful planet and learn some interesting facts regarding the exceptional SATURN:

Lowest density

Saturn is the only planet which has a lower density measure than the density of water. If Saturn could ever literally be dropped into water, Saturn would actually **float** on top of the water! Just like its neighboring gas giant Jupiter, Saturn is not a solid object. This planet is actually a fluid object as the very light elements which are contained in Saturn's composition do not condense in the frigidity of space.

Saturn

In spite of its huge size, Saturn weighs quite little. Its rocky core is enclosed by a deep layer of liquid metallic hydrogen. Saturn has less metallic hydrogen than Jupiter has. But Saturn contains **more ice**. Saturn's very low measure of density is only one-eighth as much as Earth's density.

Second largest planet

Saturn has a large equatorial diameter measuring 120537 km. This impressive length of equatorial diameter is second only to the giant planet Jupiter. The mass of Saturn is a huge 95 times greater than the mass of Earth. However, Saturn's substantial **mass** is only about **one third** as heavy as giant Jupiter's mass. Saturn's expansive diameter is approximately nine times longer than Earth's diameter.

Saturn

The total **surface area** of Saturn is **84 times larger** than Earth's total surface area. And 764 Earths could fit inside the entire sphere of Saturn! Despite being much more massive than Earth, Saturn's surface gravity is just 1.16 times stronger than the surface gravity on Earth.

Easy to locate

When Saturn appears in our night sky, it looks very much the same as any other bright star. Its existence in our universe has been known of by the peoples of many different cultures for thousands of years. But it was not always easily understood exactly what it was people were looking at. Saturn appears to be moving **slowly** in our sky which makes it a lot easier for us to locate and observe.

Saturn

Giovanni Cassini is often associated with the planet Saturn. Cassini was an Italian astronomer who discovered the **division** in the rings of Saturn named after him. He was the first person to study four of Saturn's moons. A NASA probe sent to Saturn was also named after Cassini.

Atmosphere is narrow

Saturn's clouds are much less colorful than the clouds of Jupiter. This is because Saturn's atmosphere contains a higher proportion of sulfur (Latin "sulphur"). This sulfur adds to Saturn's yellow or pale orange appearance. Its atmosphere is only a **narrow surface layer** when it is compared to the vast interior of Saturn. This narrow atmosphere contains about 75% hydrogen and about 25% helium.

Saturn

The three cloud layers of Saturn are found reasonably low in the planet's atmosphere. Hazes of smog can be found higher up above these cloud layers. Motions seen in the cloud patterns indicate the general weather of Saturn can be described as a **striped pattern** of winds.

Tilted significantly

Saturn’s rotational axis tilts at about 27° from a perpendicular line to its orbital plane. This significant angle of the tilt of Saturn’s axis causes the Sun to heat the northern half and the southern half of the planet quite **unequally**. This inequity of heat received from the Sun results in some very contrasting Saturn seasons. The inequity also accounts for the large range between its high and low temperatures.

Saturn

Saturn’s temperatures are very much colder than Earth’s as Saturn is so much farther from the Sun. Temperature at the top of Saturn’s clouds is about **minus** 175°C! The temperatures below these clouds are considerably less cold than the frigid temperatures at the top of the clouds.

Spectacular rings

The main glory of Saturn lies with its spectacular rings system. Saturn is likely best known for this eye-catching system of planetary rings continually circling the planet. Saturn's famous rings consist mostly of **ice particles** with a smaller quantity of rocky debris and dust. These distinctive and magnificent rings cause Saturn to be widely regarded as the Solar System's most easily recognized planet.

Saturn

Other giants Jupiter, Uranus, and Neptune also have rings orbiting their planet. But the rings orbiting Saturn are by far the most spectacular. The glorious rings of Saturn span a total distance of **more than 270,000 km**. But these celebrated rings are only a few hundred meters in thickness.

Rings found in 1659

Our earliest observations of the Saturn rings system were quite perplexing. The complication of viewing the rings was due to the fact Earth passes through the plane of Saturn's rings every few years as Saturn moved into its orbit. But then in 1659 Dutch physicist Christian Huygens solved the confusion regarding the **rings** and he correctly determined the specific geometry of Saturn's rings.

Saturn

Saturn's rings remained unique in our known Solar System for over 300 years. In 1977, very faint rings were discovered around Uranus, and later around Jupiter and Neptune. Saturn's rings are the most **extensive** and by far the most **impressive** of any planetary ring system.

Rings named alphabetically

The ring system of Saturn has intrigued astronomers since Galileo discovered them in 1610. Each of the rings has been named **alphabetically** in the order of their discovery. The three main rings, in order of their closeness to the planet, have been named C Ring, B Ring, and A Ring. The D Ring is an extremely faint ring but it is the nearest ring to the planet of all the spectacular rings in orbit around Saturn.

Saturn

The Cassini Division is the **largest gap** seen in the Saturn rings system. This Cassini Division separates the 'B Ring' and the 'A Ring.' Nearly all of Saturn's rings are reasonably close to each other. The only exception is the Cassini Division gap which measures about 4700 km.

Most oblate planet

Saturn has a large equatorial diameter of 120537 km compared to its shorter polar diameter of 108728 km. This is a very significant difference of about 10% in these two diameter lengths. These distinctly unequal diameters make Saturn's shape quite **oblate**. Saturn is visibly **flattened** at the north and south poles! No other planet has such a great variation between its equatorial diameter and its polar diameter.

Saturn

The flattened shape of Saturn is probably caused by the combination of its very **fast rotation** and its **gaseous** composition. The other large gaseous planets Jupiter, Uranus, and Neptune also have oblate and visibly flattened shapes, but they are still less oblate in their shape than Saturn is.

Orbit takes 29.5 years

Saturn's orbital speed is 9.69 km per second. But Saturn takes 29.5 years to complete one elliptical revolution around the Sun! This equals about 10,759 days, which makes Saturn a **slow mover** in our sky. Saturn rotates on its own axis faster than all other planets rotate with the exception of Jupiter. The visible features of Saturn are rotating at varying periods of rotation depending on their latitude.

Saturn

Saturn comes within 1.2 billion km of our Earth. The very impressive ringed planet comes within about 1.35 billion km of the Sun at Saturn's nearest point. And Saturn is 1.51 billion km from the Sun at is farthest point. Saturn's orbit is **9.5 times farther** from the Sun than Earth's orbit is.

Powerful winds

Saturn is blanketed with clouds that appear as faint stripes, jet streams, and storms. The planet has many different shades of yellow, brown, and grey. Winds in Saturn's upper atmosphere reach 500 meters per second in the equatorial region. In contrast, the strongest hurricane-force winds on Earth are a maximum of 110 meters per second.

Saturn

The pressure is the same kind you feel when you dive deep underwater. It is so powerful it squeezes gas into a liquid. Saturn's north pole has an interesting atmospheric feature - a six-sided jet stream. This hexagon-shaped pattern was first noticed in images from the *Voyager 1* spacecraft and has been more closely observed by the *Cassini* spacecraft since.

Saturn missions

Only four spacecraft in our entire history of space travel have made it all the way to Saturn. Our most comprehensive look at the spectacular planet came from NASA and the European Space Agency's *Cassini-Huygens* mission, which spent thirteen years exploring Saturn and its moons. By studying Saturn, we learn more about how our solar system formed and evolved.

Saturn

NASA's *Cassini* mission orbited Saturn, teaching us much of what we know about the ringed planet. *Voyager 1* and *Voyager 2* were both launched in 1977. Each spacecraft made a grand tour of the solar system before heading out of it. NASA's *Pioneer 10* and *Pioneer 11* were twin probes that travelled to outer planets.

Voyager 2 spacecraft

Uranus

Getting colder!

As you leave the beautiful planet of Saturn, your journey moves to Uranus. This very cold planet has been continually orbiting the Sun for billions of years, but Uranus was only discovered in the 18th century!

Your epic voyage through space now travels farther again from the Sun. The period of each planet's orbit of the Sun is growing longer with each more distant destination.

Uranus is the third destination on your tour of the four outer planets. These four very large planets are often collectively referred to as the "gas giants." And the two coldest planets Uranus and Neptune are sometimes jointly referred to as the "ice giants."

Try to keep yourself warm! Let's look at some of the most interesting details about the seventh major planet from the Sun, which has been named URANUS:

Discovered in 1781

English astronomer Sir William Herschel discovered Uranus on March 13, 1781. Herschel, who was also a composer and a fine musician, became one of the world's most **notable observers** in the fascinating history of modern astronomy. Uranus is the first planet to be initially viewed through the lens of a telescope. Herschel initially thought his newly discovered object may have simply been a comet.

Uranus

Herschel observed the new object's orbit revolved beyond Saturn's orbit. He also found its orbit was quite circular in shape. Herschel's "comet" was in fact a newly found **planet**. His exciting discovery was later named Uranus after the ancient god of the skies in Greek mythology.

Covered with icy clouds

The entire surface of Uranus consists of striking light turquoise icy clouds which comprise tiny crystals of **methane**. These methane crystals have frozen out of the planet's atmosphere. All of the methane clouds are the same light turquoise color and they completely cover the surface of Uranus. Far below these visible clouds are thicker cloud layers which comprise liquid water and crystals of ammonia ice.

Uranus

About 7,500 km below the visible cloud tops is an **ocean** of liquid water which contains dissolved ammonia. Uranus has a rocky central core which is roughly the same size as the entire sphere of our Earth. Scientists doubt very much Uranus is able to sustain any form of life.

Four times larger than Earth

Uranus is physically the **third largest** of the major planets. It ranks behind giant Jupiter and spectacular Saturn. But in weight Uranus is actually the **fourth most massive** planet in our Solar System because it is has a lower weight than Neptune which is slightly smaller in size. Uranus was never recognized as being a planet by observers in ancient times because of its dimness and very slow orbit.

Uranus

The diameter of Uranus is just over **four** times greater than Earth's diameter. But the surface area of Uranus is nearly 16 times larger than Earth's total surface area. The gravity on Uranus is only 91% of the gravity on Earth. Uranus has an equatorial diameter of about 51,118 km.

Orbit takes 84 years

The orbit of Uranus takes about 84 Earth years to complete just one revolution of the Sun! Just like every other planet in our Solar System, Uranus does not follow a perfectly circular shaped orbital path. This means Uranus has a distinctly **elliptical** path around the Sun. This path has resulted in a very significant difference between Uranus' nearest point to the Sun and the planet's farthest point from the Sun.

Uranus

A planet's nearest point to the Sun is known as its **perihelion**, which is 2.75 billion km for Uranus. The most distant point from the Sun of an orbit is called **aphelion**, and this is over 3 billion km for Uranus. Both its perihelion and aphelion occur once during its long 84-year orbit.

Rotates from east to west

Uranus rotates from east to west in the same way Venus rotates. Most planets are continually spinning on their axis in a counterclockwise direction from west to east as they orbit the Sun. The two exceptions to this direction of spin are **Uranus** and **Venus.** These two planets both rotate in a **clockwise** direction. The complete rotation period for the interior of Uranus is 17 hours and 14 minutes.

Uranus

Uranus has very **strong winds** up to 720 km per hour which blow in the same direction of the planet's rotation. Fierce winds cause some Uranus latitudes to actually rotate **quicker** than the planet's interior rotates. The latitude two-thirds from the Uranus equator to the pole has a full rotation taking just 14 hours.

Rotates on its side

Uranus is tilted so far. Its rotational axis is moving practically level with its orbital path around the Sun. Uranus has an exceptionally **high axial tilt** of 98°. This means Uranus is actually tilted at such an angle more than a right angle! The rotational axis of Uranus has basically rolled over on to its side so this axis has then become roughly parallel with the plane of the Sun. Earth is tilted about 23.5°.

Uranus

The 98° axial tilt gives Uranus **seasonal changes** completely unlike those of all other planets. Each pole on Uranus receives around 42 years of continuous sunlight! This extremely long period of sunlight is then followed by continual darkness for 42 years during the 84-year orbit.

Rings discovered in 1977

Nine narrow **rings** encircling Uranus were discovered by astronomers in 1977. These rings are each surrounded by belts of fine dust. In order of proximity to its primary planet Uranus, these nine major rings have been named as follows: 6, 5, 4, Alpha, Beta, Eta, Gamma, Delta, and Epsilon. Another two rings were discovered in 1986 by NASA's Voyager 2 spacecraft.

Uranus

Two more outer rings were discovered by the Hubble Space Telescope in 2003-2005. So the ring system of Uranus is now known to consist of 13 rings. All of these 13 rings are very **dark** and very **faint**. All except two rings of Uranus are extremely narrow. They are usually only a few kilometers wide. The rings are probably quite young.

Dynamic troposphere

The troposphere is the lowest of the three layers of the Uranus atmosphere. This troposphere is much denser than both the stratosphere and the thermosphere layers. This lowest atmosphere layer has decreasing temperatures with its higher altitudes. And the troposphere is a really **dynamic section** displaying **strong zonal winds, bright methane clouds** and some very noticeable seasonal changes.

Uranus

The troposphere of Uranus has a fairly complex cloud structure. Uranus is a world covered in **clouds**! Water clouds are below a layer of ammonium hydrosulfide clouds. Above all of these clouds are thin visible methane clouds in the uppermost region which is the coldest.

Highest proportion of ices

The extent of any warming done by the Sun is severely reduced by the fact Uranus is more than 19 times farther from the Sun as our Earth is from the Sun. The average temperature on Uranus is an incredibly hostile **minus** 197°C. Uranus does not have a conventional solid surface like our Earth has. The extremely cold atmosphere of Uranus has the highest proportion of **ices** of any major planet.

Uranus

The average distance from Uranus to the Sun is a staggering 2.87 billion km. Uranus has the **coldest** upper atmosphere of all the major planets in the Solar System. The temperature in this upper atmosphere of Uranus drops all the way down to an extremely frigid **minus** 224°C!

One spacecraft visit

The unmanned ***Voyager 2*** is the only spacecraft which has ever visited Uranus. NASA's *Voyager 2* was launched from Earth in August 1977 and has been flying around our Solar System and beyond ever since. Its pioneering itinerary took the spacecraft to all four outer planets. *Voyager 2* flew past Uranus in January **1986**.

Uranus

Voyager 2 spacecraft came within about 81,500 km of the extremely cold planet. Over many years, thousands of images of Uranus have been radioed back to Earth from the *Voyager 2* spacecraft. This space probe has transmitted a huge amount of data back to us regarding the planet's numerous moons, its rings, its atmosphere, and the whole Uranus environment.

Twin spacecraft

NASA launched twin spacecraft *Voyager 1* and *Voyager 2* in 1977. They were both designed to last about five years and to conduct a study of Jupiter and Saturn, the larger moons of these two planets, and Saturn's spectacular rings. But the missions continued to go on and on with flybys of the outermost planets Uranus and Neptune.

Uranus

The two Voyager spacecraft are the third and fourth probes to fly beyond all outer planets in our Solar System. *Pioneer 10* and *Pioneer 11* preceded Voyager in leaving the Sun's gravitational attraction. The closest approach to Uranus by *Voyager 2* occurred in 1986. *Voyager 2* is still the only probe to visit **Uranus**. The Voyagers revolutionized planetary astronomy.

Recent discoveries

In 2005, NASA announced the discovery of a new pair of rings around Uranus and two new, small moons orbiting the planet. The finds were made from photos taken by the Hubble Space Telescope. The largest ring discovered by Hubble is twice the size of the previously known rings. In 2011, NASA's *New Horizons* spacecraft crossed the orbit of Uranus on its way to Pluto.

Uranus

New Horizons became the first craft to journey past Uranus' orbit since *Voyager 2*. The spacecraft did not take images of Uranus. The possibility of a flagship mission to Uranus will be a focus of planetary science at NASA in the years to come. *New Horizons* will observe Uranus from its location in the outer solar system.

Voyager 2 over Neptune

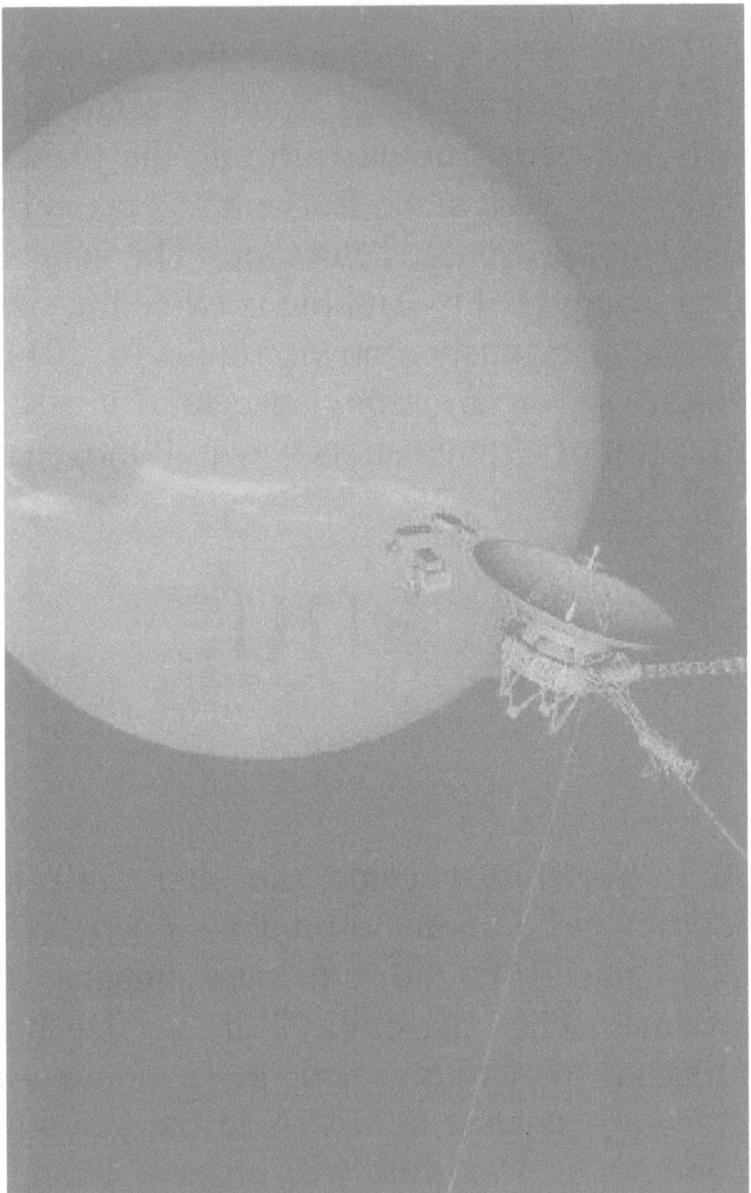

Neptune

Final frontier

Your amazing journey through our Solar System has covered an unimaginable distance of about 4.5 billion kilometers! Remember the Sun was your starting point for this amazing astronomic adventure.

The Sun is located at the central nucleus of our Solar System. But you are now approaching the very cold and the very remote outer areas of our Solar System. You are now moving further out from Uranus.

You have traversed billions of kilometers to now reach the eighth major planet. This voyage through outer space has taken you to some absolutely fascinating destinations. Each of the eight planets has their unique and interesting qualities.

Your final major planet also has some unique features. Let's now explore the highlights and very interesting characteristics regarding the very distant NEPTUNE:

Discovered in 1846

Before 1846, the French mathematician named Urbain Le Verrier made a really bold prediction a new planet would be discovered in a specific section of our sky. And then ultimately in September 1846, Neptune became known as the **most distant** planet in our Solar System after it was seen in the predicted area of the sky. Berlin astronomer Johann Gottfried Galle used calculations by Urbain Le Verrier.

Neptune

Pluto was discovered in 1930, and named the most distant planet. But there are periods when Pluto orbits the Sun on a path nearer to the Sun than Neptune is. In 2006, it was decided Pluto is **not** actually a planet! So Neptune yet again became the most distant planet in our Solar System!

Fourth largest planet

Neptune is substantially **larger** than each of the four **inner** terrestrial planets Earth, Venus, Mars, and Mercury. Neptune's equatorial diameter measures 50,539 km. Its volume is so expansive you could place 57 planet Earths inside Neptune's whole sphere. The total surface area of Neptune is about 15 times larger than the total surface area of Earth. But from our viewpoint far away on Earth, Neptune appears as a just a dim and distant object in our night sky.

Neptune

Neptune is seen as a tiny blue-green dot in our sky. The tiny dot's slow movement can be tracked in our night sky over a period of weeks and months. In terms of **physical** size, Neptune is the **smallest** of the **four giant** outer planets. These four are collectively referred to as the **gas giants**.

Made of ices

Average distance between the outermost major planet Neptune and our Sun is nearly 4.5 billion km! At the top of its clouds, Neptune's temperature goes all the way down to a freezing minus 218°C! The composition of really cold Neptune is predominantly **ices** such as water ice, ammonia ice, and methane ice. Deep in the interior of Neptune the composition is primarily pressurized ices and rock.

Neptune

Neptune has an extremely active climate with extremely high winds which speed around the planet faster than the planet itself rotates! Its atmosphere is composed primarily of hydrogen and helium. The two extremely frigid planets of Neptune and Uranus have sometimes jointly been referred to as the **"ice giants."**

Strongest winds

Neptune has a huge storm in its southern hemisphere called the **Great Dark Spot**. This prominent feature is about half the size of Jupiter's Great Red Spot. Mighty winds can rip across Neptune at 2100 km per hour! Neptune has the strongest sustained winds ever recorded on any planet. The Great Dark Spot is dark blue in color. Most winds on Neptune blow in a westward direction.

Neptune

Neptune's other major storm system is called the **Scooter**. This storm is further south than the Great Dark Spot and it circumnavigates the planet of Neptune roughly every 16 hours. The possibility of life on Neptune is extremely unlikely due to the incredibly high wind speeds.

Differential rotation

As Neptune is not a solid body, its gassy atmosphere has **differential rotation**. The wide equatorial zone of Neptune rotates fully with a period of over 18 hours. By stark contrast, the remarkably quicker rotation period of Neptune's north and south poles is only about 12 hours. This very considerable differential rotation of Neptune's latitudes is clearly the most pronounced of any major planet.

Neptune

It is estimated the rotation period of Neptune's inner core is 16 hours and 7 minutes. Neptune's sphere appears **oblate** in its shape with a **bulge** at the **equator** and flattened poles. Neptune is not a perfect sphere as its equatorial diameter is about 900 km longer than its polar diameter.

Season lasts decades

Neptune takes 164.8 years to orbit the Sun! This is by far the **longest** period of orbit of any major planet. This orbital period is actually longer than your whole life will be! This is not particularly surprising if you consider the enormous distance between the path of Neptune's orbit and the Sun. On 29 May 2011, Neptune completed exactly **one orbit** of the Sun **since** the time Neptune was discovered in 1846.

Neptune

Just like Earth, Neptune has four seasons. But each Neptunian season will continue for longer than 41 years! Neptune is continually spinning on its rotational axis which is always tilted toward the Sun at a 29° angle. This **axial tilt** is responsible for the change of seasons.

Appears blue

The methane gas on Neptune gives the planet its distinctive **azure**, blue color. This is because methane gas absorbs red light and reflects blue light back into space. **Methane** will **absorb** the longer visible wavelengths of **red** and **yellow light** so the reflected sunlight appears blue. Neptune appears a noticeably much bluer color than Uranus because Neptune has **more methane** in its atmosphere.

Neptune

Methane accounts for less than 5% of Neptune's total atmosphere. Helium and molecular hydrogen comprise the vast majority of the Neptune atmosphere as they do on all of the four largest planets. Neptune's atmosphere comprises clouds of frozen methane ice crystals.

Five rings

Four of Neptune's four rings are narrow. And the other ring named Galle is broad, but it is more diffuse. The dark coloring of this broad ring is due to chemical reactions of highly charged particles striking material in the rings. All of these five rings are comprised of **dust particles** produced by the many meteorites which have smashed into one of Neptune's moons.

Neptune

The rings are named: Galle, Le Verrier, Lassell, Arago, and Adams. Small moons orbiting Neptune **shepherd** the narrow rings and keep them from expanding. Neptune's rings are fainter and darker than the rings of Saturn. The outermost Neptune ring Adams is unique as it has three curved arcs denser than the rest of the ring.

Enormous pressure

Neptune has the fastest winds to have ever been recorded in our Solar System. But the extreme **energy** for the rapid winds is not coming from the Sun. Neptune radiates 2.6 times more energy than it absorbs from the Sun. Like Jupiter and Saturn, Neptune seems to have an internal heat source. The enormous **pressure** deep inside Neptune builds up until gaseous hydrogen turns into liquid metallic hydrogen.

Neptune

As the pressure builds, the methane in the Neptune atmosphere also decomposes as bonds **dissolve**. It is possible carbon atoms bind to each other in this extreme pressure environment to form diamonds. Nobody is certain, but a rain of diamonds may be falling towards Neptune's core!

First visit in 1989

The **final** planetary target for NASA's spacecraft *Voyager 2* was Neptune. In August 1989 *Voyager 2* passed just 4,950 km above the northern pole of Neptune. This was the **nearest** approach to any planet since NASA's *Voyager 2* space probe departed our Earth twelve years earlier. *Voyager 2* continued to move at a fast average velocity of 19 kilometers per second for twelve years to reach Neptune.

Neptune

Our Solar System's most distant major planet is thirty times farther from the Sun than our Earth is! The NASA space probe continued to observe planet Neptune for another four months. *Voyager 2* continues to explore and it is now heading out of our Solar System towards **interstellar** space.

Dim Neptune

Neptune is so far from the Sun that high noon on the big blue planet would seem like dim twilight to us. The warm light we see here on our home planet is roughly 900 times as bright as sunlight on Neptune. As it is so far from the Sun, it's the only planet in our solar system that is not visible to the naked eye. In 2011 Neptune completed its first 165-year orbit since discovery in 1846.

Neptune

Sometimes Neptune is farther from the Sun than dwarf planet Pluto is. The eccentric and oval-shaped orbit places Pluto inside Neptune's orbit for a 20-year period every 248 Earth years. But Pluto can never crash into Neptune because for every three laps Neptune takes around the Sun, Pluto makes two. The repeating pattern prevents a crash.

Deep atmosphere

Neptune does not have a solid surface. Its atmosphere is made up mostly of hydrogen, helium, and methane. This atmosphere extends to great depths, gradually merging into water and other melted ices over a heavier, solid core with about the same mass as Earth. Scientists think there might be an ocean of super-hot water under Neptune's cold clouds.

Neptune

The water does not boil away because incredibly high pressure keeps it locked inside. It seems contradictory to view an ice giant like Neptune with a superheated ocean under those blue clouds. But that's what makes our solar system fascinating. We are constantly learning new and unexpected facts about our neighbors.

Apollo 11 to the Moon

The first manned lunar mission was launched from Kennedy Space Center

Planet's moons

Fly me to the Moon!

You have run out of the Solar System's major planets! But a bonus tour now awaits you. Let's explore all the moons orbiting their primary planets.

People have been fascinated with the changing phases of Earth's Moon through the ages. But most of the moons observed in orbit around other major Solar System planets have only been discovered by astronomers in fairly recent times.

Six of the eight major planets have moons constantly orbiting their primary planet. Our Earth has just one natural satellite which we simply refer to as the "Moon." Five of the major planets have multiple moons.

Let's now find out exactly how many known moons each planet has orbiting them. And let's appreciate the unique qualities of each moon and look at some of the most notable MOONS located in our Solar System:

Our nearest neighbor

The distance between our Earth and the Earth's Moon is an average of about 384000 km. This great distance between Earth and Moon is about 30 times longer than our planet's diameter. Our Moon measures only about one quarter the length of Earth's diameter. Our Moon has a gravity just one sixth as strong as Earth's gravity. Our Moon is the **fifth** largest moon in the entire Solar System.

Planet's moons

Our Sun's diameter is about **400 times** longer than the diameter of our Moon. Coincidentally, our Sun is nearly **400 times** farther from Earth than our Moon is. This coincidence means the Sun and the Moon **appear** to be fairly **similar** in **size** when they are viewed by us from Earth.

Motion of the ocean

Gravity keeps our feet on the ground and it is this gravity which also keeps our Moon trapped in its continual orbit around planet Earth. And gravitational forces are also responsible for the rise and the fall of our ocean's tides, with respect to the land, all around our wonderful planet. The two principal agents affecting these ongoing tidal rise and fall motions of the world's oceans are our **Sun** and our **Moon**.

Planet's moons

Our Sun is nearly 400 times farther from Earth than our Moon is from Earth. So the smaller, but much nearer, Moon has a stronger effect on us than the Sun has. Moon's **gravitational** forces are strong enough to disrupt our water balance by **accelerating** water towards the Moon.

No Mercury or Venus moon

Both Venus and Mercury have no moons or natural satellites. Earth has only one natural satellite (named the Moon) whilst Mars has two moons orbiting the planet. All four of the **outer** planets Jupiter, Saturn, Uranus, and Neptune have **numerous moons** continually orbiting their primary planet. Venus and Mercury do not have a moon orbiting because both of these inner planets are positioned **too close** to the **Sun**.

Planet's moons

Any moon which may be orbiting too far from either Mercury or Venus would be in an unstable orbit and would be captured by the Sun. But if any moon were too close to the inner planets of Mercury or Venus, this moon would then literally be destroyed by tidal gravitational forces.

Mars has two moons

Phobos and **Deimos** are two small moons orbiting above the equator of Mars. These two moons are both massive lumps of rock about the size of an average city! Phobos is the only satellite which orbits its primary planet faster than the planet itself rotates. Phobos is the larger of the two moons. Both of these small moons were discovered in 1877 and both of them have a highly **non-spherical** shape.

Planet's moons

Deimos has a fast orbital period which nearly matches the period of rotation of Mars. Deimos rises as expected in the **east**, but it rises very slowly. Despite the 30-hour orbital period of Deimos, this small moon takes 2.7 days to ultimately set in the **west** as Deimos gradually falls behind the slightly faster rotation of planet Mars.

Jupiter has 95 moons

Scientists keep discovering more moons orbiting Jupiter! Most of the 95 known moons are relatively small. The four largest Jupiter moons are referred to as the **Galilean satellites.** These four bright moons were found by Italian astronomer Galileo Galilei in 1610. The four moons are named: Io, Europa, Callisto, and Ganymede. Callisto is the most cratered moon in the entire Solar System.

Planet's moons

Discovery of the four Galilean satellites was the first very key evidence Earth was **not** actually at the center of the whole universe. Galileo was able to show these four moons were in **orbit** around **Jupiter.** This was the proof he needed to show **not** all of the objects located in the sky were orbiting our planet Earth.

Ganymede is largest moon

Ganymede is the **largest moon or satellite** found anywhere in the entire Solar System. Looking outward from Jupiter, Ganymede is the seventh moon orbiting the Solar System's largest planet. This extremely large moon is composed of silicate rock and water ice. The remarkably long diameter of Ganymede is actually longer than the diameter of the Solar System's innermost small planet Mercury!

Planet's moons

Ganymede orbits Jupiter at a distance of over one million km. Its orbital period is 7 days and 3 hours. Ganymede is **tidally locked,** which means one side always faces Jupiter. For every one Ganymede orbit, Jupiter's moon Europa orbits the giant planet twice as Io orbits Jupiter four times.

Saturn has 274 moons

The first discovery of a moon orbiting Saturn occurred in 1655. And by 1997 Saturn's moon count had grown to 18. Space probes sent to study Saturn in recent years found numerous moons and increased the Saturn moon count to **274 known moons** with confirmed orbits. Each Saturn moon is unique and it tells another piece of the fascinating story regarding the history of the whole Saturn system.

Planet's moons

Saturn's Titan is the second largest moon in the entire Solar System. Titan measures 5150 km across and it is unique among moons as it has a **dense** atmosphere. Titan's thick air has organic smog in it. The dense Titan atmosphere comprises about 95% nitrogen with traces of methane.

Uranus has 28 moons

Most of the 28 known moons orbiting Uranus have been named after characters from the works of the famous playwright William Shakespeare. And a couple of the Uranus moons are named after the works of Alexander Pope. The planet of Uranus is about ten thousand times more massive than all of its 28 known moons. All of the moons orbiting Uranus are comprised of **ice** and **rock**.

Planet's moons

Titania is the largest moon which orbits Uranus. It is the eighth largest moon in the entire Solar System. But Titania is about twenty times less massive than our Earth's Moon. Titania was discovered by Sir William Herschel in 1787, six years after he discovered the planet of Uranus.

Neptune has 16 moons

All of Neptune's sixteen known moons are composed of rock and ice. Triton is the only moon which orbits in the **opposite** direction to the direction Neptune spins. The four innermost moons named Naiad, Thalassa, Despina, and Galatea are so close to Neptune they actually orbit the planet within Neptune's ring system. The large moon Triton comprises 99.5% of the mass known to orbit Neptune.

Planet's moons

The surface of Neptune's moon Triton is very **active** with geysers erupting liquid nitrogen. Triton's lowest ever recorded temperature is an uncomfortably chilling **minus** 235°C. This is by far the **coldest** temperature ever recorded anywhere in the entire Solar System!

Man's footprint on our Moon

First successful mission to the Moon

Man on the Moon!

July 20th, 1969, is the celebrated date a man first walked on our one and only Moon. Astronauts flew toward the Moon in 1959, but the monumental 1969 walk on the Moon's surface ended a very keen **space race** between USA and the Soviet Union. Our Moon remains the only other extraterrestrial body to be visited by humans - so far! NASA's Apollo 11 flight was launched from Florida in USA.

Planet's moons

Apollo 11 was crewed by Neil Armstrong, Edwin "Buzz" Aldrin, and Michael Collins. Our first manned landing on the Moon was successfully made when Armstrong and Aldrin landed their lunar module, named Eagle, on the surface and then soon became the **first** people to **walk** on the Moon!

Home sweet home

Astronauts always remark how lovely Earth looks from space, especially when compared to the barren surface of the Moon. Our planet is a wonderful place which provides us with the ideal conditions to support life – and our lifestyle!

You have now completed your journey of the planets in this Best Practice book. Well done! But don't end your journey here! You have studied such an extremely tiny portion of the entire universe.

This book is your ideal introduction to the planets. Hopefully, this book has increased your appetite to digest more and to explore worlds beyond our Solar System. You will continue to learn in your endless search of an incredibly vast and stimulating universe.

Astronomy is a fascinating study of a wonderful universe. What we have learnt so far about our impressive Solar System is astronomical! What objects are out there just waiting to be found? Watch this space!

Rating the planets:

Largest planet	**JUPITER**
Smallest planet	**MERCURY**
Fastest spinning planet	**JUPITER**
Slowest spinning planet	**VENUS**
Hottest planet	**VENUS**
Coldest atmosphere	**URANUS**
Highest peak	**MARS**
Most volcanoes	**VENUS**
Largest canyon	**MARS**
Fastest orbit of the Sun	**MERCURY**
Slowest orbit of the Sun	**NEPTUNE**
Highest density	**EARTH**
Lowest density	**SATURN**
Highest axial tilt	**URANUS**
Most oblate shape	**SATURN**
Largest storm	**JUPITER**
Strongest winds	**NEPTUNE**
Most moons	**SATURN**
Most impressive rings	**SATURN**
Best surf	**EARTH**

In the same way you are a unique person in the universe, each planet is also unique. You are a really special individual! You are one of a kind! This is also the case with the planets of our vast universe as no two planets are exactly the same.

Where to next?

Beyond Neptune's orbit is a huge **belt** of icy bodies. This outer belt is even larger than the Asteroid Belt which orbits the Sun between the orbits of Mars and Jupiter.

The very distant region beyond Neptune is known as the **Kuiper Belt**, named after Dutch astronomer Gerard Kuiper. Over one thousand Kuiper Belt objects have been discovered recently but many more objects as yet undiscovered are believed to exist.

Any object in the Solar System orbiting the Sun at a greater distance on average than Neptune orbits the Sun is defined as a trans-Neptunian object (TNO). The largest TNOs are designated as **dwarf planets.**

In order of size, the four dwarf planets are Eris, Pluto, Haumea, and Makemake. Pluto has a long orbital period of 248 years! The larger Eris is our Solar System's largest known dwarf planet. Eris was only recently discovered in 2005 but we know it takes 557 years to complete am orbit of the Sun!

Exoplanets

An exoplanet is any planet which is beyond our Solar System. Most exoplanets are orbiting other stars in our universe.

Most of the exoplanets discovered so far are in a relatively small region of our galaxy, the Milky Way. We know from NASA's Kepler Space Telescope there are more planets than stars in our galaxy.

The composition of exoplanets ranges from rocky (like Earth and Venus) to gas-rich (like Jupiter and Saturn). Exoplanets are made up of elements similar to those of the planets in our Solar System, but their mixes of those elements may differ. Some planets may be dominated by water or ice, while others are dominated by iron or carbon.

Exoplanets come in a wide variety of sizes. The first exoplanets were discovered in the 1990s using a variety of detection methods. We live in a universe of exoplanets. The count of confirmed exoplanets is now over 5860; and this number is rising fast.

Resources

Books:
Solar System Observer's Guide - Peter Grego
Patrick Moore's Astronomy - Sir Patrick Moore
The Universe – Collins gem

Websites:
www.nasa.gov
www.universetoday.com
www.solarviews.com
voyager.jpl.nasa.gov
www.nineplanets.org

Popular Environmental Science websites

usda.gov
scielo.org.mx
earth.com
ecologiaverde.com
udocz.com

Books in series:

Be Better

Live better, plan better, learn better, look better, love better, eat better, work better, and much more! This is your personal **growth** bible.

New Testament nuggets

Read the greatest collection of spiritual verses written in the 1st century. All 27 New Testament books are represented with profound verses.

Universe tour

Take the **ultimate journey** through our vast universe. Visit planets, moons, stars, and galaxies. Travel to the most distant galaxy.

Bible highlights

Take a **tour** and follow the Bible with selected verses. Learn **valuable** lessons from 112 verses, including the world's favourite Bible verse.

French tongue

French is presented in an easy-to-follow format. Learn essential words and phrases. And learn the correct pronunciation of words and phrases.

Born in 20th century

All quotes are from people born in the 20th century. More than a thousand profound events are listed. One hundred years in your hands!

Books in series (2)

Acts, Romans, Corinthians

Acts, Romans & Corinthians are the sequel to the Gospels. Luke and Paul wrote these books because Christians wanted to know Jesus more.

Confucius says

The profound quotes of Confucius affect one quarter of humanity and still echo in the mind. All these sayings are universal and timeless.

Italian tongue

The passionate language of Italian is presented for you in an easy-to-follow format. Prepare for a trip to Italy – the holiday of a lifetime.

Jesus says

Jesus is the most influential person in all of human history. Adopt these best practices in your life, as taught by the son of God, Jesus.

Math essentials

Come on a journey from numbers and symbols to the various branches of mathematics. Build your math skills one practice at a time.

Psalms & Proverbs

Psalms and Proverbs are the “core” of the Bible. Profound verses give valuable life lessons. Read more of the Old - a Testament full of gold!

Books in series (3)

New You

Want to build a new you? Develop a new plan, gain new energy, and build a new mind, body, and spirit. Create and nurture a new you.

Tennis essentials

Practice ball sense and footwork to build tennis skills for a lifetime of **fun** and **fitness**. Learn the **essentials**. Practice the **shots**. Play the **game**!

OLD into NEW

The New Testament is our divinely inspired commentary on the Old. It's beneficial for us to see how authors of the New understood the Old.

Echoes of wisdom

This is a fascinating compilation of inspired quotes from the greats of our modern history. A treasure trove of timeless quotes from our past!

Spanish tongue

Spanish is the main language in 21 countries. Learn essential words and phrases. Practice the correct pronunciation of words and phrases.

She said

Over two hundred women share inspirational quotes with you. Profound quotes come from Cleopatra to the Queen, to Greta Thunberg.

Books in series (4)

German tongue

German is the **most spoken** language within the European Union. Learn essential words and phrases and use their correct pronunciation.

Winning

What is winning? What are the qualities of a winner? How do you experience winning more often? Success is an issue of character, not performance. Read the ground-breaking book before your opponent reads it!

"I am"

Affirmations help to inspire and motivate you. Your "I am" affirmations encourage you to be your best self. They are a great way to build the mentality and qualities of a winner.

Greek greats

Philosophy is a Greek creation. Ancient Greek philosophy was used to make sense of the world using reason. It's still relevant now. Learn from the profound quotes of 25 outstanding thinkers.

Five words or six words

Short quotes are packed with wisdom. Quotes help you reveal and assess your assumptions, your values, and your beliefs. Five words to motivate you. Six words to change your life.

Save our planet

We are blessed with a beautiful planet. We used to be surrounded by greenery, water, mountains, desert, and snow. But now Earth is crowded with buildings and millions of homes in urban areas.

There are many things you can do to help protect our planet Earth. Youcan start by cutting down on what you throw away. Follow the three “R’s” to conserve batural resources and landfill space:

Reduce
Reuse
Recycle

When you further your own education, you can help others understand the importance and value of our natural resources. You can also volunteer for cleanups in your community.

Trees provide food and energy. Planting trees will help save energy. Trees help to clean the air and help to combat climate change. Planting trees also helps to prevent soil erosion. Tree planting is one of the most effective ways of tackling climate change caused by greenhouse gas.

One day we will be gone. But do you have the realization of what we are giving to our kids?

Vision Statement

Best Practice is the ultimate way to learn valuable practices about life.

Grow your **knowledge** and **confidence** with best practice learning and progression.

www.ingramcontent.com/pod-product-compliance
Lightning Source LLC
LaVergne TN
LVHW090409210926
848480LV00002B/514